Behalal Yopson Antoine Cesar

Seismic stratigraphy, log interpretation

Behalal Yopson Antoine Cesar

Seismic stratigraphy, log interpretation

and 3D modelling of the reservoirs in the SW part of the RIO DEL REY offshore/Cameroon Basin

ScienciaScripts

Imprint

Any brand names and product names mentioned in this book are subject to trademark, brand or patent protection and are trademarks or registered trademarks of their respective holders. The use of brand names, product names, common names, trade names, product descriptions etc. even without a particular marking in this work is in no way to be construed to mean that such names may be regarded as unrestricted in respect of trademark and brand protection legislation and could thus be used by anyone.

Cover image: www.ingimage.com

This book is a translation from the original published under ISBN 978-620-2-28128-7.

Publisher:
Sciencia Scripts
is a trademark of
Dodo Books Indian Ocean Ltd. and OmniScriptum S.R.L publishing group

120 High Road, East Finchley, London, N2 9ED, United Kingdom
Str. Armeneasca 28/1, office 1, Chisinau MD-2012, Republic of Moldova, Europe
Printed at: see last page
ISBN: 978-620-5-87280-2

TABLE OF CONTENTS

INTRODUCTION

The geological history of the Rio Del Rey sedimentary basin has been known since the 1980s, and the majority of the large oil fields discovered more than 20 years ago are already tending to be depleted. According to the latest report declared by the national oil company, production in Cameroon has dropped from 26.8 million barrels for an average daily production of 73,094 barrels in 2009 to 21.4 million barrels in 2011 for 63,211 barrels per day, a decrease of 13.08% compared to 2008. Faced with this considerable drop in production and according to estimates made in a report presented by the firm Moore Stephens in 2012, the reserves would be in full decline to the point of affecting the country's economy. These results call into question the potential of the current aging fields and therefore open the way for new explorations and a re-evaluation of oil deposits. However, this will allow us to hope for a reversal of the trend in the coming years when we know that the only truly productive basin in Cameroon is the Rio Del Rey basin, which covers about 7,000 km^2 of offshore area and provides nearly 9/10ème of the national crude oil production from 55 fields.

The purpose of this work will be to show that in this basin, there are still hydrocarbon reserves that could probably be exploited and that could fill the deficit observed during the previous years, especially from 2008 to 2011. Thus, it will be based on the evaluation of a block located in the southwest of the Rio Del Rey from 3D seismic data and wells. This study will consist of determining the type of trap, the place of accumulation of hydrocarbons, the extension of a reservoir and the estimate in volume of its content by designing a 3D model that is as close to reality as possible. In order to achieve the above objectives, the work was organized in three chapters:

- **Chapter I Generalities on the RDR basin**: this will be the place to present the RDR basin, from a geological point of view, thanks to a synthesis of previous studies.

- **Chapter II Materials and methods used for a subsurface analysis**: in this part we will present the methodology and data that will allow us to achieve the important results, in order to achieve the objectives set.

- **Chapter III Results and discussion:** we will try to propose an interpretation of the seismic results, the well results and the volumetry. We will also compare the results obtained with those of neighbouring fields and finally provide a critique with respect to the amount of data available.

PRESENTATION OF THE LABORATORY

1- LOCATION AND DESCRIPTION

The laboratory of petroleum geosciences is located at the University of Douala under the Department of Geosciences. It was created on March 22, 2011 following a cooperation between the university, the national company of hydrocarbons (SNH) and the multinational Schlumberger specialized in the oil industry. The latter has made available to the university ten (10) licenses of PETREL 2010 software for 10 workstations, each with two screens connected to a central unit with a hard disk of 800 gigabytes and a memory of 8 gigabytes, the machines being connected to a server.

This work-friendly research laboratory (Fig.1) facilitates the professional training of future petroleum geologists through its accessibility and all its comfort and safety facilities.

2- OPTIONS

The laboratory is run by senior researchers from the university, professionals from different oil companies in the area as well as partners such as SNH and Schlumberger who provide the logistics and expertise necessary for the training of young petroleum geologists.

3- THE MISSIONS

The Petroleum Geosciences Computational Laboratory serves as the research framework for the Professional Master's Degree in Petroleum Systems Geosciences. The main mission of the laboratory and its Master is to train young geologists and geophysicists from the petroleum industry in the methods and techniques of exploration, evaluation and characterization of new fields or reservoirs likely to contain hydrocarbons (oil and gas).

The specific objectives are to train high-level specialists capable of pursuing their careers in doctoral schools in various fields of research such as geology, geophysics, geochemistry, transport in porous media, computer science and numerical modelling, and also to enable candidates wishing to enter the world of industry to develop the skills necessary for the professions of reservoir geologist, exploration geologist and applied geophysicist.

4-   OPERATION

The functioning of the laboratory is linked to that of the professional Master in Petroleum Systems Geosciences. The training takes place in one year divided into two semesters. The courses are constituted of UE (Teaching Unit). Each UE is composed of theoretical courses.

Fig. 1: Partial view of the Petroleum Geosciences Computer Laboratory

CHAPTER I . GENERALITIES ON THE RIO DEL REY BASIN

In Cameroon, 95% of the oil production is based in the Rio Del Rey Basin. However, its exploration and development is more recent than in the Douala Basin, which currently constitutes the remaining 5% of production associated with the Kribi-Campo sedimentary basin (SNH, 2009). In other words, all the oil exploration activities carried out and also most of the discoveries made in Cameroon have been for the most part carried out off the West African coast.

1.   Regional situation

The Rio Del Rey basin extends south of the Niger Delta and is located in the Gulf of Guinea. Located in the southwestern region of Cameroon, it lies between 4°43'59" north latitude and 8°39'0" east longitude. This basin constitutes the eastern terminus of the Biafra cone, one of the two deltaic cones of the large Niger Delta (Mascle, 1977). Thus, it corresponds geodynamically to the confluence of fracture systems related to both the Cretaceous Benue graben (NNE-SSW) and oceanization (general N-S and E-w orientation) (Perrodon, 1983). It is separated from the Douala Basin by the Cameroon Volcanic Line (CVL) and forms the border between Nigeria and Cameroon (Fig. 2).

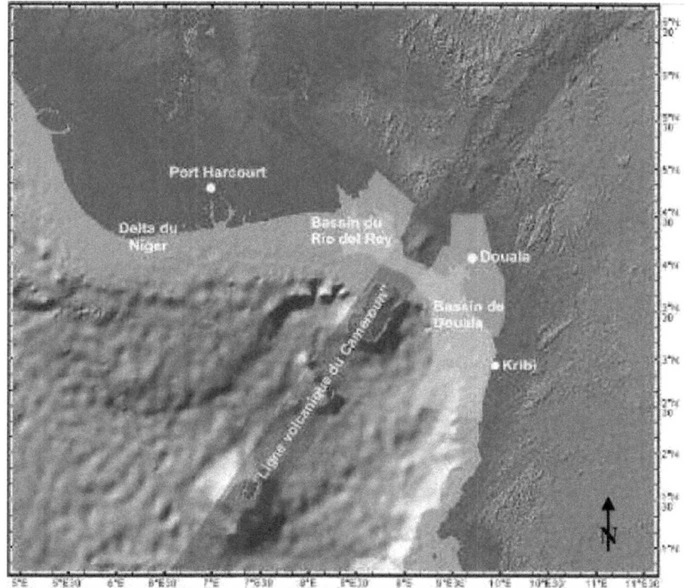

Fig. 2 Geological map of Cameroonian coastal basins: location of the Rio Del Rey (SNH/UD, 2005)

2. Structural signature

The Rio Del Rey Basin is a passive margin basin resulting from the separation of the African and South American continents following the opening of the South Atlantic (Mvondo Owono, 2010). It is considered a Tertiary age basin (Schiefelbein et al. 2000) and its tectonic-sedimentary evolution is very similar to that of the Niger from which it receives its sedimentary inputs (Mvondo Owono, 2010). The particularity of this basin is that most of the traps encountered are of tectonic type and result from argillocinesis compared to the Douala basin where most of the traps are stratigraphic. Indeed, the Rio Del Rey basin is subdivided into three structurally different petroleum provinces (Subra and Brussot, 1988):

•   North of the basin where gas has been discovered: the E-W trending Growth Fault Province (**GFP**) encompasses the coastline and mangroves; it corresponds to the middle zone of the Cross River Delta.

7

- In the center of the basin where we find the largest oil fields in Cameroon such as the Ekoundou fields, the discoveries made are essentially oil and are found in the N-S oriented clay ripple province (**PRA**) which results from overpressure phenomena in under compacted clays.

- To the south of the basin, a province of clay domes (**PDA**) that form a frontal bulge characterized by reverse faults set up following an overthrust of deltaic deposits and corresponds to the downstream area of the delta.

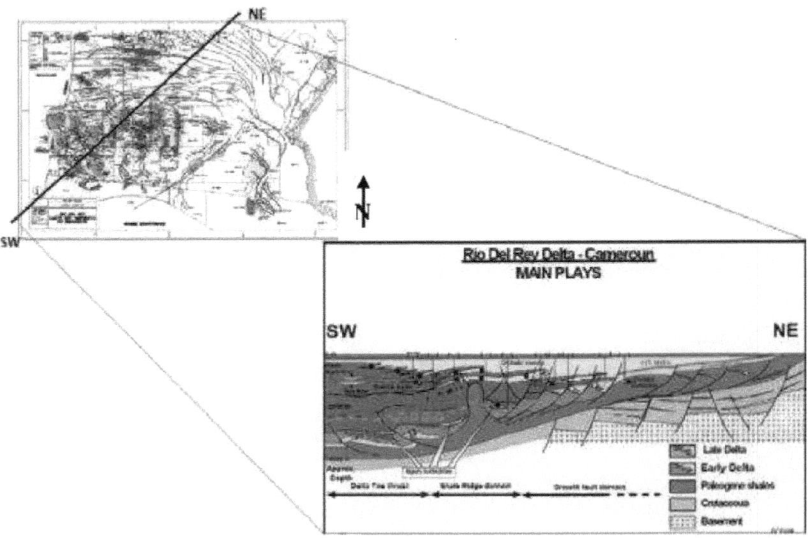

Fig. 3. structural section of the Rio Del Rey basin (modified from SNH, 2009)

This structuring is due to gravity tectonics with the extension upstream (north) and the compression corresponding to the damping downstream (south).

3. Stratigraphic and lithostratigraphic presentation

Subdivisions were made taking into account the sequences of deposits on the geological time scale and the nature of the facies, thus defining its stratigraphy and lithostratigraphy.

a) Lithostratigraphy of the basin

The RDR is a basin with three main formations, each corresponding to specific environments. These formations include the Benin Formation, the Agbada Formation and the Akata Formation (Saugy et al. 2003).

• The Benin Formation constitutes the massive continental sands intercalated with clay beds that formed from the Upper Miocene to the present. This formation constitutes the coastal and alluvial plain.

• The Agbada Formation forms a silico-deltaic sequence intercalated with sands and silts designated by S on the RDR stratigraphic chart and clays designated by M and which also date from the Upper Miocene to the present (Benkhelil et al. 2002). It is found in the delta front.

• The Akata Formation constitutes unconsolidated marine clays that were deposited from the Paleocene to the present. Their depositional environment is the pro-delta.

b) Stratigraphy of the basin

During 40 million years, the sedimentary evolution of the Rio Del Rey margin was done in such a way that today its stratigraphy can be divided into two units according to the time scale (Subra and Brussot, 1988):

- A pre-deltaic unit from the Paleocene to the Miocene :

The clay deposits, notably marine silts and clays in deep water, sediment over a thickness of 800 m in the PFC, 1000 m in the PRA and 200 m in the PDA (Blin,

1997). These materials subsequently yield clay diapirs under the effect of lithostatic pressure. Subsequently, the slopes of the diapirs cause slumping in the poorly consolidated sediments. On the other hand, the Eocene series are generally made up of lenticular turbidites (the Oongue and Debunsha sands) of low extension and which are close to the deposits of the lower slopes (Blin, 1997). The Oligocene is marked by a major marine regression that would have caused violent submarine erosion. A similar depositional system continued to be developed during the Lower Miocene.

- **A deltaic unit from the Miocene to the present**:

The Lower Miocene is characterized by N-S oriented turbiditic deposits extending downstream for more than 100 km, sandstone deposits develop until the Middle Miocene at the point where a deep fan is established in the "ripple trench" (Blin, 1997). During the Upper Miocene, the Niger-Cross River Delta front spreads widely over the entire downstream sector, covering the clay diapirs of the PRA, but not reaching those of the Eastern Province (EP). This progression results in a displacement of the upstream zone of the delta, with notably a great extension of the sands in the diapirs zone which globally form prograding depositional sequences deposited essentially during the end of the Miocene and the beginning of the Pliocene marked by some transgressive pulses from the Upper Miocene to the present (Le Dluz and Montagnier, 1996).

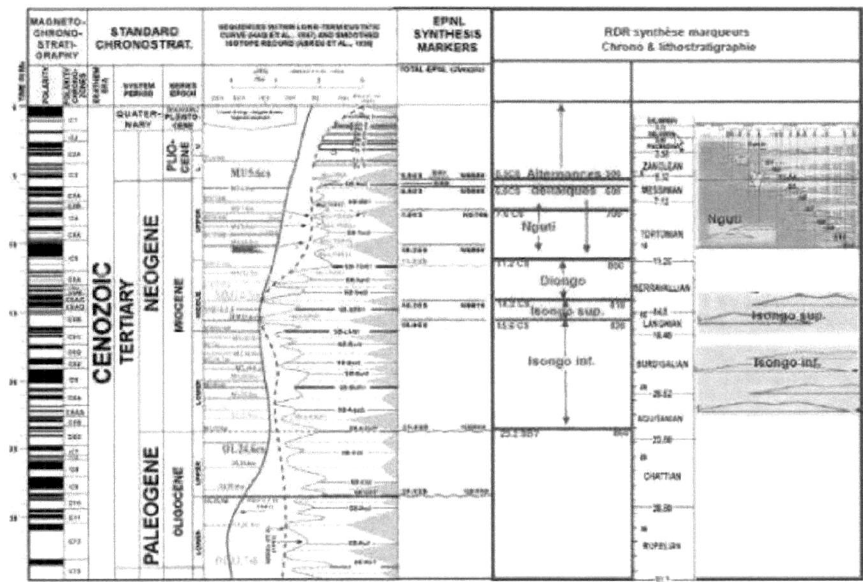

Fig. 4. stratigraphic chart of the Rio Del Rey basin (SNH, 2005)

CHAPTER II. MATERIALS AND METHODS USED FOR SUBSURFACE ANALYSIS

The study of a block located in the southwestern Rio Del Rey is based particularly on petroleum data obtained from seismic acquisition and wells provided by the Department of Geosciences of the University of Douala. The interpretation of these data is based on the sequence stratigraphy that allows to define the stratigraphic units that are the expression of three parameters: tectonics, sedimentary flow, eustatism.

A- MATERIALS

The data used are seismic lines and logs, mainly delayed logs. Due to their confidential nature, they will be processed without any precise indication of their geographical location.

- **Seismic data**

The data acquired in 1980 consist of 752 3D seismic lines in dip and 1340 in strike of poor quality and low resolution covering the entire study area of approximately 155.1 km2. The dip seismic profiles (perpendicular to the shoreline) are oriented E-W and 16.5 km long, while the strike seismic sections (parallel to the shoreline) are oriented N-S and 9.4 km long. The seismic used will allow us to identify facies units and define the traps present in the subsurface. We have plotted only the seismic lines used for seismic interpretation (Fig. 5).

- **Well data**

In addition to the seismic lines, we have data from two exploration wells, P1 and P2, at depths of 1500 m and 1360 m respectively. These wells provide conventional logs such as natural radioactivity (GR), resistivity (ILD), spontaneous potential (SP), neutron density (RHOB and NPHI), well diameter (CALIPER) as well as checkshots (velocity data) obtained from the travel time of a sonic wave in a rock (Dt) Four types of logging measurements will be used in this study: Gamma ray (GR), this is the radioactivity emitted naturally by a formation, its tool measures the percentage of radioactive elements contained in a formation, resistivity (ILD) which is the ability of a fluid to resist the passage of electric current, Neutron density logs (NPHI and RHOB) which are nuclear logs because their measurement principle is based on the excitation of atoms of a formation by a bombardment either by gamma radiation (RHOB) or by neutrons (NPHI).

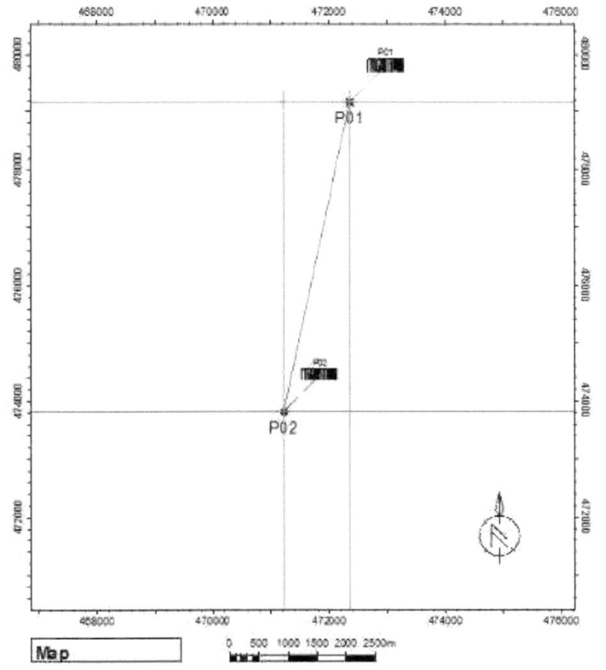

L1A and L1B: Seismic lines passing through P1

L2A and L2B: Seismic lines passing through P2

L3: Seismic lines passing through P1 and P2

Fig. 5. layout of the wells and seismic lines studied (offshore)

Tools

This study was done using PETREL 2010 software in conjunction with EXCEL 2007 and ADOBE Illustrator. Indeed, PETREL is a software for the interpretation of 2D and/or 3D seismic data, well data and reservoir modeling (*see methodology on PETREL*).

8- METHODS

The objective of this work is to show from the data that have been assigned to us that in this study area, there may be other hydrocarbon reservoirs with a non-negligible volumetric estimate. And for this, we used as methods: seismic stratigraphy, qualitative and sequence analysis applied to logs, modeling and volumetry from PETREL 2010.

a) Seismic interpretation procedure (seismic stratigraphy)

Seismic stratigraphy is a method of interpreting seismic data to reconstruct the geometry of sedimentary bodies, to identify the sedimentary processes that contributed to their emplacement, and to extract qualitative information about the lithology (AAPG, 1977). A seismic interpretation contributes to the development of a sedimentary model and depends on information extracted from a two-scale analysis:

> At the scale of the seismic section

It consists of an analysis of seismic discontinuities and an analysis of seismic facies units.

Analysis of seismic discontinuities corresponds to the identification, pointing, and geological interpretation of the remarkable seismic horizon located through geometric relationships with underlying and overlying reflectors (onlap, downlap, toplap) and are interpreted in terms of stratigraphic discontinuities as sequence boundaries, transgression surfaces, and maximum flooding surfaces (Mitchum and Vail, 1977) (Fig. 6)

Seismic facies analysis which is done in 3 steps: definition of seismic facies units, their mapping and interpretation.

The definition of seismic facies units, which consists of characterizing a group of seismic reflections whose configuration, external shape and internal parameters (continuity, amplitude, frequency) differ from adjacent units (AAPG, 1997). It allows the definition of a number of seismic facies.

Seismic facies mapping allows the identification or delineation of different seismic facies.

The interpretation of the seismic facies consists in giving the characteristics of the different mapped facies in terms of lithology, depositional environments and sedimentation rates.

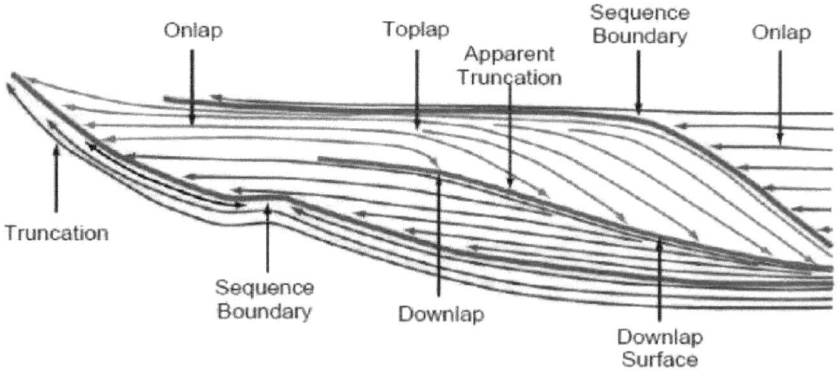

Fig. 6. processions identified by reflectors (P. Vail, 1988)

> On the scale of seismic reflection

The analysis is based on the characteristics of seismic reflectors such as continuity and amplitude variations. The continuity translates the influence of lithological parameters (argillosity, porosity, fluid saturation) on the shape of the seismic traces while the amplitude variations predict and quantify, from the basin scale to the reservoir scale, the characteristics of the sedimentary filling such as, for example, the basement defined by blurred reflectors with hyperbolas or brightspots (very high amplitudes) which indicate a potential hydrocarbon content.

b) Log interpretation procedure

Logs are records of petrophysical parameters of subsurface rocks during or after drilling. The purpose was to use a qualitative "quick look" analysis to determine the lithology of the formations but also their fluid content.

> Quick look analysis

This is a rapid log interpretation method that consists in determining the lithology of the formations crossed, the sequences of sedimentary deposits, the presence of hydrocarbons and the contact between fluids in the formation from the signatures of the logs obtained during or after drilling. Indeed, the standard logs used are :

- The gamma ray (GR) makes it possible to distinguish clay formations containing many radioactive elements (U, Th, K) from formations that do not contain them such as sands (Segesman, 1980). Thus, the GR will increase in clay formations and decrease in sand formations (NB: this does not apply for radioactive sands such as feldspathic sandstones which require the spectral gamma ray log).

- The resistivity (ILD) is an absolute parameter in the determination of hydrocarbons. Indeed, the presence of hydrocarbons in reservoirs is reflected by a very high resistivity. However, a high resistivity does not necessarily indicate that we are in a hydrocarbon reservoir because it may be fresh water, it should therefore integrate all other logs to identify the quality of fluid present in the formation.

- The neutron density logs (NPHI and RHOB) are also indicators of the presence of hydrocarbons. A combination of the two logs indicates that we are in oil, while a combination indicates that it is gas. On the other hand, they are also used to determine the lithology of the formations crossed in the sense that a crossing of the two logs will indicate a change of facies, in particular a sandstone reservoir if RHOB on the left and NPHI on the right (negative polarity) or clay if the trend is reversed (positive polarity). On the other hand, if there is an overlap between the two (zero polarity), it will be limestone. For a good interpretation, it would be necessary to integrate at the same time all the logs of the various petrophysical parameters.

-

c) sequence analysis applied to logs

Sequence analysis applied to the logs aims at determining the sequences of deposits, which will allow to make correlations in order to better judge the extension of a defined reservoir. It is based on the analysis of the shape of the logs such as the GR from which are identified the positive sequences (decreasing granulometry), the negative sequences (increasing granulometry). Indeed, according to the shape of the log, we can distinguish several particular forms closely related to the depositional environment:

- The funnel-shaped forms that reflect prograding deposits that are put in place when the energy of the environment increases (Bourquin, 1991), this in the case of marine regression. These deposits can be found in a fluviodeltaic environment.

- The bell-shaped forms which translate retrograde deposits found for example on the platform, when the energy of the environment decreases during a marine transgression.

- Cylinder shapes that indicate aggrading deposits emplaced in a quiet environment with generally uniform sedimentation and may be encountered in the basin proper.

However, these forms can be jagged (sawtooth) which marks the alternations of fine sand and clay banks.

d) Procedure used for modeling and volumetry (PETREL 2010)

The realization of this study was conceived on 5 steps far to know: the importation of the data, the mapping of the surfaces, the shimming well-seismic, the modeling in 3D and the calculation of the volumes.

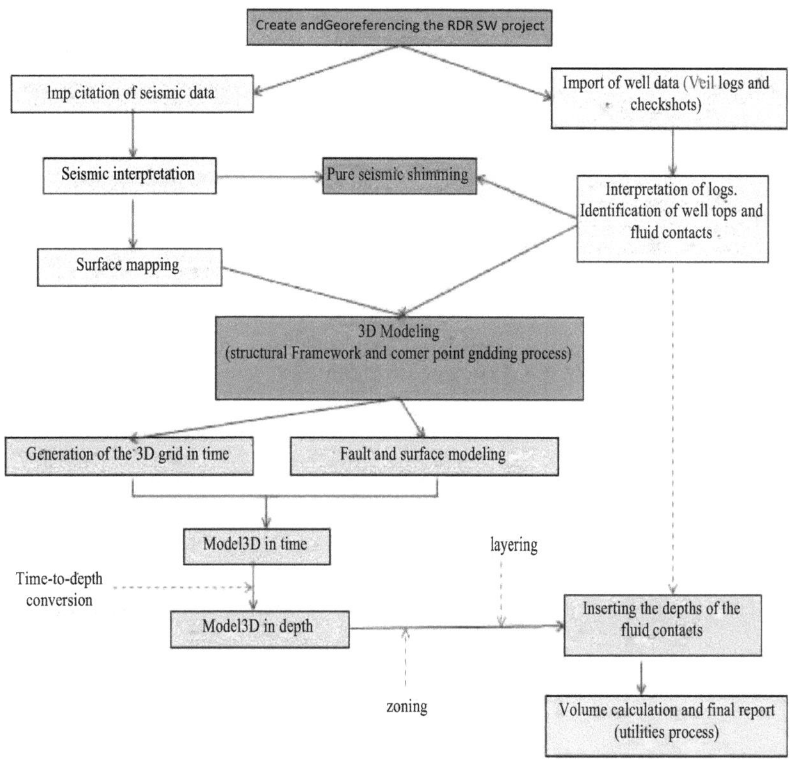

Fig. 7: PETREL 2010 usage diagram

CHAPTER III . RESULTS AND DISCUSSION

The results of the present study are explained in this chapter. The results of the analysis of the seismic profiles, the wells, the modeling and the volumetric results are presented and discussed later.

A- RESULTS

1- Seismic facies analysis and trap characterization

The seismic lines L1A and L1B are the ones that pass through the well P1 as long as the seismic lines L2A and L2B pass through the well P2.

a) Analysis of seismic facies

The L1A and L2A dip seismic lines have 3 major facies units depending on the configuration of the seismic reflectors, their shapes and internal parameters. These are:

The summit unit. It is a stratified zone with more or less parallel and moderately continuous reflectors of very strong amplitude and whose external shape indicates draping sheets. It is probably deposits of sands constituting the Benin Formation that have not undergone compaction.

The intermediate unit. This is a stratified zone, the reflectors are mostly parallel to subparallel and very continuous in places. The amplitude of the reflectors is medium to low. These are probably alternating clays and low-energy sands in deep water that were put in place when subsidence was regular and the sedimentation rate uniform.

The basal unit. It probably corresponds to the Akata clays dating from the Paleogene. It is an acoustic basement that can be assimilated to chaotic reflectors (absence of geometrical configuration). These are probably clays deposited in a low energy environment (fine clayey sediments) during a transgression phase.

Each unit has several facies bounded by discontinuities. We have identified seven discontinuities:

> dl, d3, d4 and d7: these are discontinuities that could be transgressive surfaces because they are characterized by reflectors that abut them in onlap and toplap. Discontinuities dl and d3 constitute the limits of the Sl depositional sequence while d3 and d4 constitute those of the S2 depositional sequence. The S3 depositional sequence observed only on the L2A seismic line is bounded at the base by d4 and at the top by d7 which does not appear on the LlA seismic line.

> d2 and d6: these are probably erosional surfaces characterized by erosional truncations on which the reflectors come to rest in downlap and toplap. The erosional surface d6 is the boundary between the Benin Formation and the deltaic alternations, and appears on both seismic profiles at about 500 s depth. As for the d2 surface, it is found between 1250 and 1500 s depth only on L2A.

> d5: this is a discontinuity that could be assimilated to a maximum flooding surface. It is characterized by reflectors that abut it in downlap and is located between d6 and d4 only on the seismic line LlA.

Some of the facies identified are shown in the table below:

Table 1: Characterization of facies

Seismic facies	description		samples	lithology	Sedimentary interpretation
Lick A	stratified zone, more or less parallel reflectors and moderately continuous, strong amplitude, draping sheets			Probably from the sands of Benin	Sedimentary deposits in a low to medium energy environment, uniform sedimentation rate and regular subsidence
Facies B	Stratified zone, parallel to subparallel reflectors of medium to low amplitude and arranged in parallel sheets	discontinuous		Probably clay and sand deposits in deep water	Sedimentary deposits from a low energy environment to a high energy environment (calm to agitated)
Lick C		Very continuous			
Facies D					
Facies E	Acoustic base that can be assimilated to chaotic reflectors of variable amplitude.			under compacted clays	Low energy deposits (fine and clayey sediments)

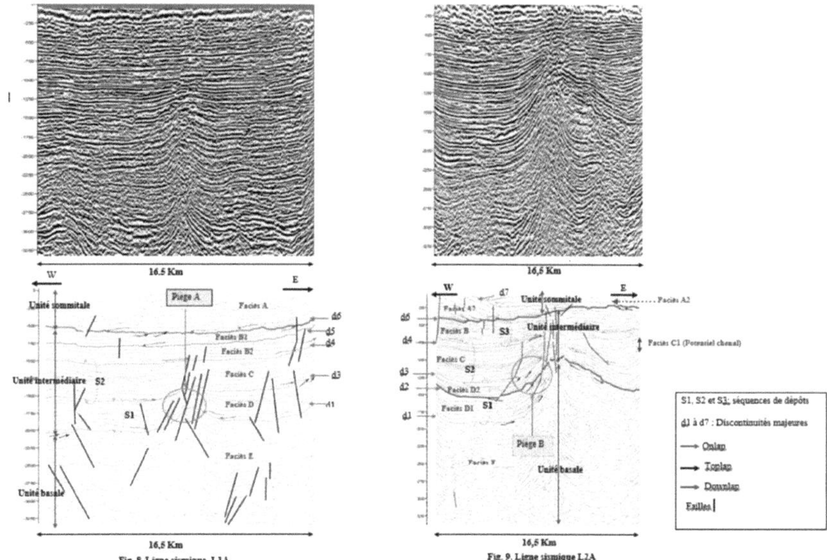

Fig. 8. Ligne sismique L1A

Fig. 9. Ligne sismique L2A

According to the analysis made, there could have been brief transgressive phases that favored the deposition of Akata clays in deep water in a calm energy environment. There were also mainly prograding and aggrading deposits of high marine level in the clay-sand alternations constituting the intermediate unit and finally a great erosion followed by prograding deposits of sands constituting the Benin Formation.

b)   Trap characterization

The highly fractured clay rift zone has an abundance of faults on the seismic lines in dip (L1A and L1B) and strike (L2A and L2B). Indeed, traps A and B are located in alternating clays and sands where we find traps closed in 3 dimensions by normal and post-sedimentary faults. This structuring results from an intumescence of undercompacted clays because they are filled with pore waters playing the role of a detachment level (Fig.10 and 11).

23

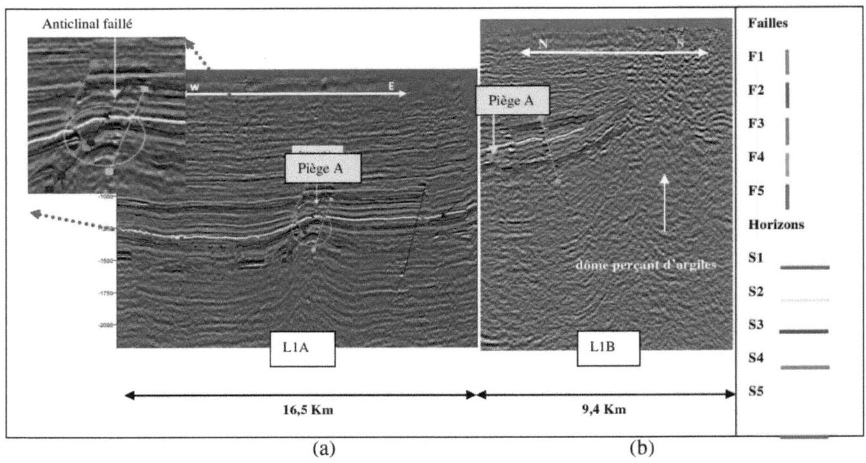

Fig. 10. Characterization of trap A. (a). Dip structure. (b). Strike structure

The potential reservoirs are closed at the dip and strike by normal faults and at the tops by probable clay covers. These faults act as a barrier to fluids such as hydrocarbons, which can be detected by brightspots or flat spots.

Trap A is defined on the LIA seismic line as an al faulted anticline with a clay core (Fig. 10 (a)) and on the L1B seismic line as a faulted closure (Fig. 10 (b)).

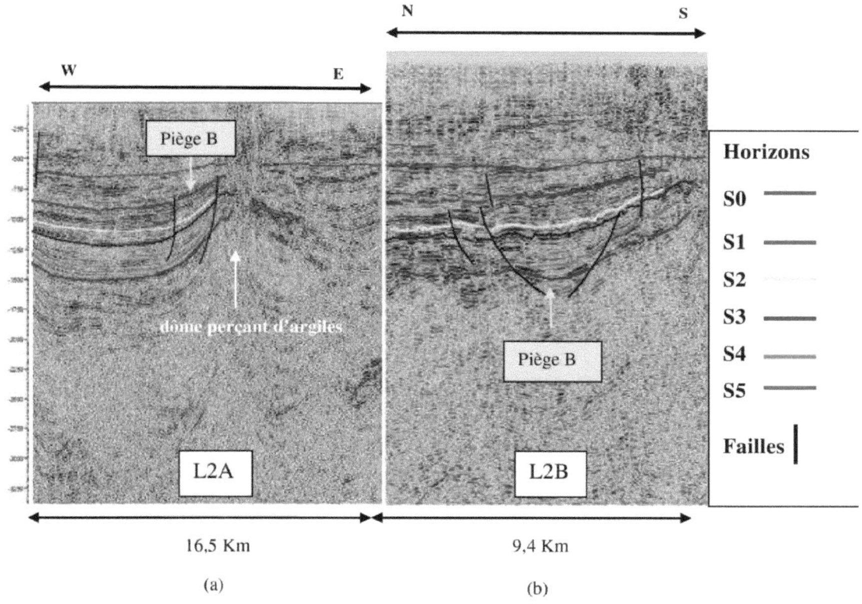

Fig. 11. Characterization of trap B. (a). Dip structure. (b). Strike structure

Trap B is defined by two N-S oriented normal and strike-slip faults that dip to the W, one of which is bordering the western flank of the clay dome (Fig.11 (a)). These faults constitute a N-S barrier to potential reservoirs and are due to gravity tectonics and the piercing dome.

The seismic line in strike L2B (Fig.11 (b)) shows a trap closed by two antithetic listric faults on the northern flank of the dome. Indeed, these two E-W oriented normal and strike-slip faults dip in opposite directions, one to the south and the other to the north.

The defined traps are consequently closed in dip and strike by boundary faults at the flanks of the dome.

c)   Structural map and location of traps A and B

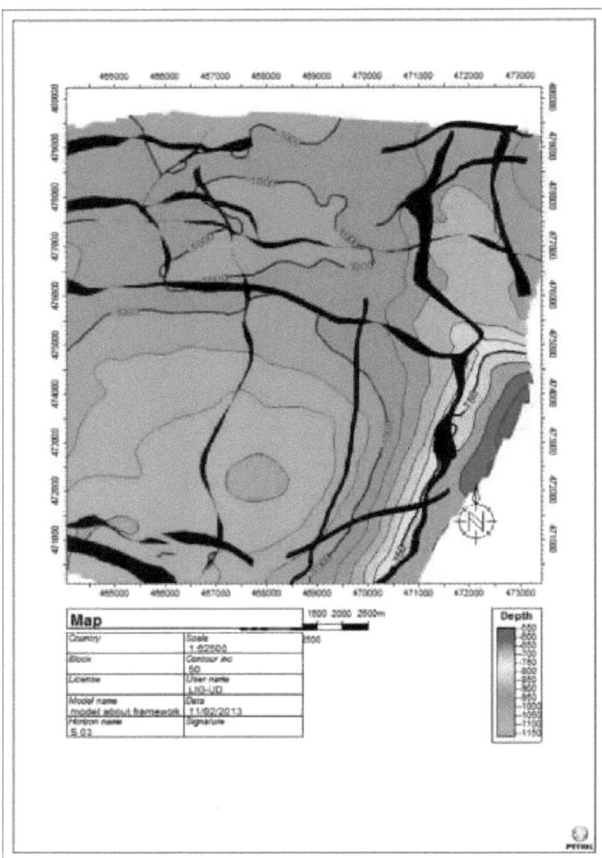

Fig. 12. structural map S4

The above map shows high and low zones. The high zones (potential hydrocarbon accumulation zones) are limited by a large boundary fault on the W side of the clay dome. The low zones correspond to the synclines.

Trap A, located at the northern end of the study area, is framed by two large N-S oriented faults and two small E-W oriented faults.

Trap B lies a little further south in the area where the clay intumescence is much more pronounced and is also closed by two N-S faults and two other E-W trending faults.

2. Presentation of well logs from P1 and P2 (Fig. 13)

Two exploratory wells targeting deltaic alternations have provided information on the lithology of the formations crossed and on the nature of the fluids they contain. Thus, well P1 targets trap A located between 1000 m and 1200 m depth while well P2 targets trap B located between 1000 m and 1500 m depth. The generalities are listed in the table below.

For this study, core and cuttings data are missing!

Table 2: General information on wells P1 and P2

Names of the wells	Wellhead coordinates in X	Wellhead coordinates in Y	KB	Top MD	Bottom MD	Units
P1	---------------	---------------	20.1	0	1500	m
P2	---------------	---------------	21.3	0	1360	m

KB: height of the rotation table from ground level

Top MD: reference surface from which the drilling starts

Bottom MD: depth of borehole measured from the drill string

The parameters used in the composite log are gamma ray (GR), spontaneous potential (SP), resistivity (ILD) and density-neutron logs (RHOB and NPHI).

- Qualitative analysis (lithology and fluid content)

The reading of the GR at 900 m depth allows to identify on the left of the cutoff placed at 40 the sand banks (probably sandstone) and on the right the clay banks. These are sands mixed with a considerable percentage of clays that attenuate the GR peaks to the left of the cutoff.

In comparison with the RHOB and NPHI, many crossings indicate a change from sandy to clayey facies indicating that we are in the deltaic alternations.

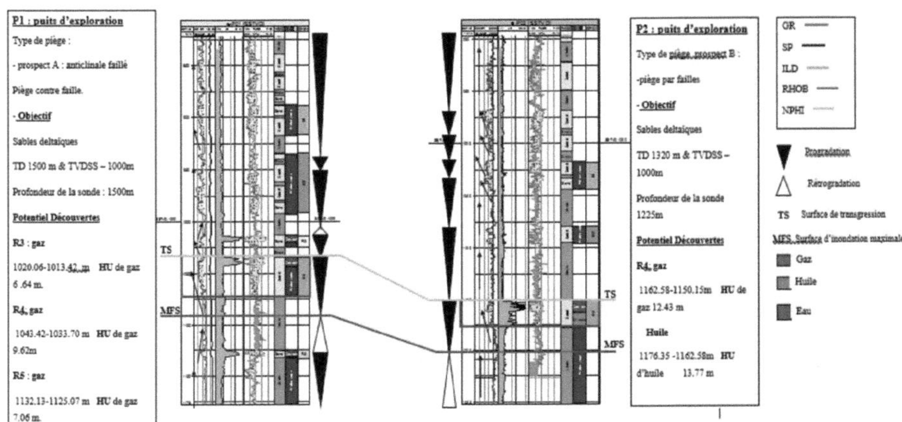

Fizure. 13. Presentation of wells P1 and P2

Several reservoirs are defined, including R1, R2, R3, R4 and R5 at a depth of over 1000 m. The zones considered interesting are those with high resistivities, with peaks observed at depths of 1020 m, 1132 m, 1043 m and 1162.58 m for both wells, indicating the presence of hydrocarbons, particularly gas for reservoirs R3, R4 and R5 in well P1 and oil for reservoir R4 in well P2. In addition, the R1 and R2 reservoirs contain salt water characterized by a low resistivity value.

In the R4 reservoir observed at 1162.58 m depth, the oil-gas contact is marked by the boundary between very high and medium resistivity peaks as well as an approximation of the RHOB (left) and NPHI (right).

- **Sequential analysis**

An analysis from the GR log indicates two main types of forms: cylinder and funnel forms, which imply sequences of prograding and aggrading deposits along the depositional profile. Their upper contact is abrupt, marking an abrupt change from sandy to clayey facies. These forms are saw-toothed, probably related to tidal and swell movements, causing fine laminations of sands and clays.

During the progradation, the sedimentation rate is quite important and the coarse sediments coming from the continent are deposited on the fine clayey sediments and constitute granocroissant sequences at the origin of these funnel shapes which are mainly found in the clayey-sandstone series and which constitute in this particular case the deltaic alternations. The progradation is followed by an aggradation with a rather important sedimentary flow. There was thus a succession of prograding sequences linked to a marine regression that set up the sandy reservoirs R1, R2 and R4. These reservoirs can be separated from the clays above by a synchronous marine transgression surface that could constitute in this case a reference (datum) for the elaboration of genetic units.

- **Deposit environment**

The depositional sequences identified from these forms are therefore mostly prograding and aggrading, indicating the possibility that these sediments were deposited in an environment strongly influenced by eustatic variations, particularly marine regression and submarine currents. Given the geological context in which we find ourselves, i.e. a fluvio-deltaic environment, it is more likely that it is the deltaic front or sub-aquatic deltaic plain located in the neritic zone. This environment is characterized by alternating prograding sandy deposits of the Agbada with the Akata clays and in which the reservoirs from R1 to R5 are found.

3. Calibration and correlation of the R4 tank

The R4 reservoir located at a depth of 1150.15 m (MD) in the P2 well is of petroleum interest with a considerable oil height (13.77 m). We will follow its lateral evolution by correlations in order to see its extension in subsurface.

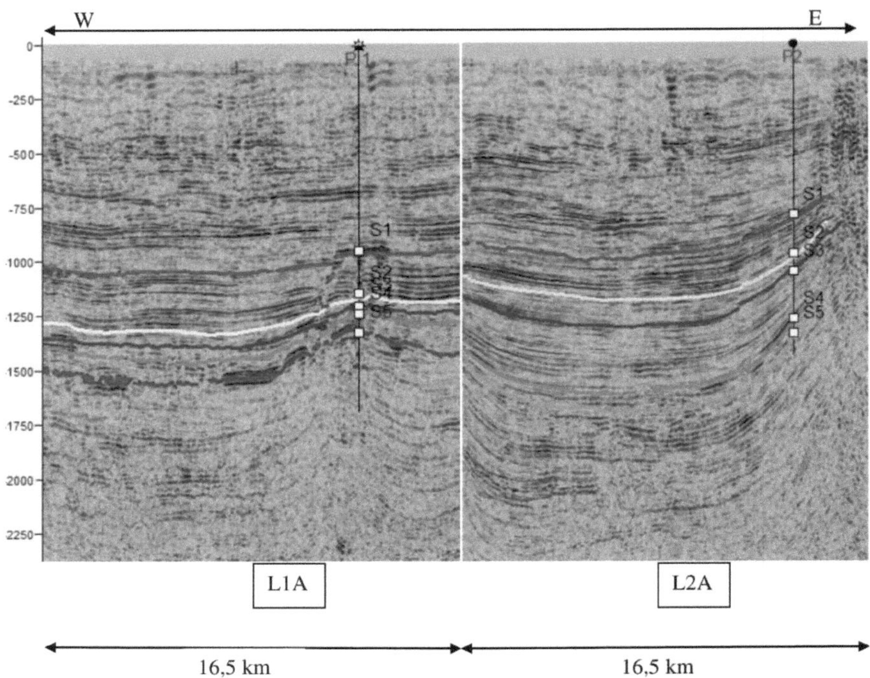

Fig. 14. Seismic well padding

The correlation of the R4 reservoir was done in part through the logging of each well, the calibration and reconstruction of the depositional architecture of the sandy sediments of the R4 reservoir. This was done through the tracking of reflectors but also with the depositional sequences. Indeed, the fact that we are in a fluvio-deltaic environment, the prograding sandy reservoirs deposited in a marine regression context are separated from the clays above by a synchronous marine transgression surface (TS).

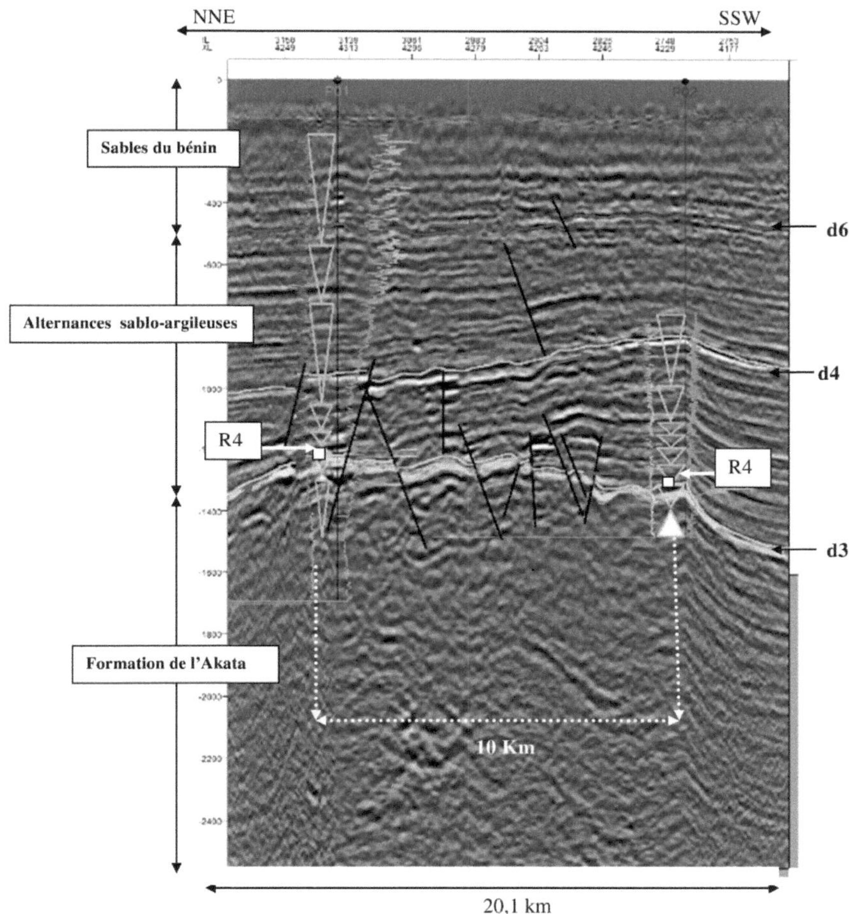

Fig. 15. Correlation of the R4 tank on the L3 seismic line

d4 and d3: transgression surfaces

d6 : sequence limit

Reflector

Rifts

The R4 sandy reservoir deposited in the context of marine regression extends for about 10 km along well P1 to well P2. However, there are many post-sedimentary faults between these two wells that could have acted as a barrier to other fluids. With respect to the reservoir content, there was a secondary migration. Indeed, the hydrocarbons would have migrated from the low zones to the high zones inside the reservoir where they were stopped by tight faults. The oil and gas migrated to the SSW of the R4 reservoir to accumulate in the upper part where the P2 well is located while the gas accumulated to the NNE of the reservoir.

4. Isobath maps, oil maps and 3D model

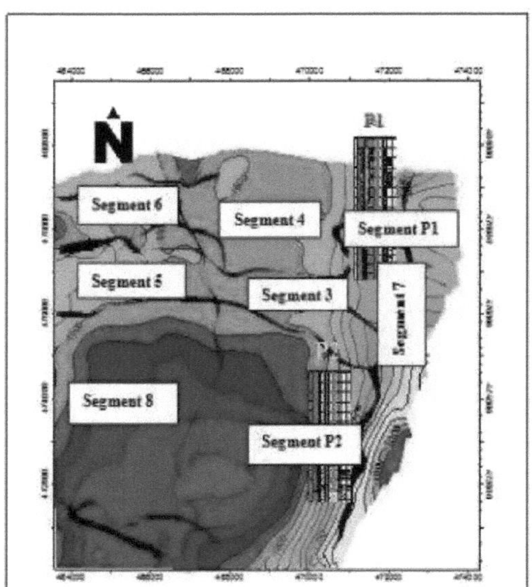

Fig. 16. S4 maps showing the different segments

The structural map above (Fig. 16) shows the layout of the faults throughout the study area as well as the position of the two exploration wells P1 and P2. From the 3D modeling on Petrel, these faults were connected to each other at the point where segments were created. Each segment being delimited by faults, we have 8 segments within which a volume of hydrocarbon will be determined.

The surface map S4 (Fig. 17) presenting the R4 reservoir shows that the hydrocarbons have effectively migrated from the lower parts (shallows) which can be valleys at about 1500 m depth to the high zones at about 1160 m and have accumulated laterally along the clay dome and limited by a large N-S trending boundary fault. This reservoir therefore contains oil and gas and is differentiated by 3 contacts namely:

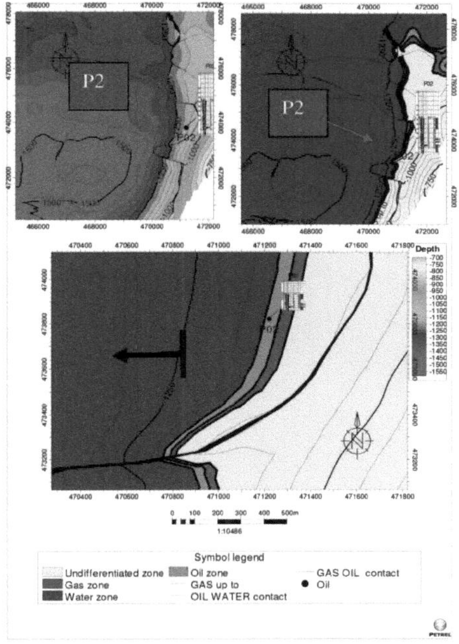

Fig. 17. Isobath and oil map of the S4 surface

-      The contact between the gas and a clay formation (GUT: gas up to) located at 1150.15 m depth.

-      The contact between oil and gas (GOC: gas oil contact) located at 1162.58 m depth.

-      The oil-water contact (OWC) is located at a depth of 1176.35 m.

The 3D model obtained from the 5 generated surfaces and the thickness maps allow to differentiate 4 zones among which: the zone 4 included between the surface S4 and S5 inside which is the tank R4 (Fig. 18).

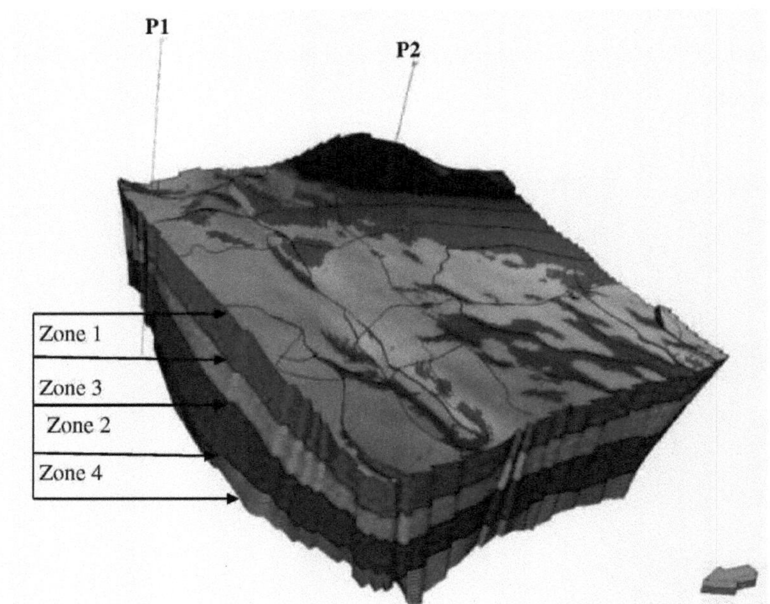

Fig. 18. 3D modeling results

5. Hydrocarbon volume ratio

General property : porosity = 0.34000 Net/Gross = 0.8000

TABLE 3: Parameter values used for volumetry

Bg (training vol. factor)	Rv (vaporized oil/gas ratio)	Gas recovery factor	Sw	So	Sg	Bo (training vol. factor)	Rs (solution gas/oil ratio)	Oil recovery factor
1.0000	0 .0000	1.0000000	0.300	1-Sw-Sg	0.3000	1.210000	0.10000	1.000000

TABLE 4: volumetric results

VOLUMER 4	BULK VOLUME (10*6 m3)	NET VOLUME (10*6 m3)	PORE VOLUME (10*6 rm3)	HCPV OIL (10*6 rm3)	HCPV GAS (10*6 rm3)	STOIIP (in oïl) (10*6 sm3)	STOIIP (in gas) (10*6 sm3)	
	1570	1256	427	112	103	112	103	
ZONE 4 (TR4- BR4)	1565	1252	426	71	103	71	103	
ZONE 4 (BR4- TR5)	5	4	1	41	0	41	0	
Segment P1	238	190	65	0	37	0	37	
Segment P2	708	583	254	28	24	28	24	
Segment 3	59	82	35	31	17	31	17	37
Segment 4	129	103	22	0	0	0	0	0
Segment 5	29	7	7	0	0	0	0	20
Segment 6	73	58	20	0	0	0	0	0
Segment 7	32	226	21	53	25	53	25	0
Segment 8	302	1	3	0	0	0	0	0

11	0	0	0	11	0
5	0	0	0	5	0
0	0	0	0	0	0

STOIIP = quantity in place in the tanks at initial conditions (stock tank initially in place)

HCPV = hydrocarbon pore volume

BULK VOLUME = tank volume

NET VOLUME = volume of the part of the tank containing only hydrocarbons

PORE VOLUME = volume of interconnected pores making up the reservoir

NET/GROSS = ratio between the volume of the part containing the hydrocarbons and the total volume of the Tank

B - DISCUSSION OF THE RESULTS

1) Result of the seismic analysis of the facies

Since we do not have core or cuttings data, the facies could not be dated. Comparing this result to those obtained in the work done by Dr. MVONDO OWONO.F in his thesis (Cenozoic Surrection of the West African Passive Margins, 2010), we note similarities between the facies units he dated in the basin to those obtained in the SW RDR. Thus, we believe that the seismic facies analysis of the study area presents three major units that can be equated respectively from the bottom to the top with the Paleocene-aged Akata Formation (basal unit), the Agbada-Akata sandy-clay alternations (intermediate unit), and the Miocene to present-day Benin Formation (summit unit). Thus, the great erosion limit (d6) separates the Benin Formation of the present day from the sandy-clay alternations of the upper Miocene and the Pliocene. Consequently, at the end of the Pliocene there was a cessation of sedimentation followed by a strong erosion in the basin. The resumption of sedimentation was made by deposits of prograding sands constituting the Benin formation.

2)   Volumetric results for R4 tanks

The steps to follow throughout this exploration phase lead to a conclusive result in terms of evaluation of the SW part of the Rio Del Rey from a block. Nevertheless, it is necessary to take into account the uncertainties on the value of saturation, porosity and resistivity of the fluids, which cast doubt on the reliability of the results obtained. Nevertheless, although these results are purely estimated, we have highlighted the presence of several hydrocarbon reservoirs, notably R4, whose estimated oil volume (approximately 176 million barrels) appears to be close to that obtained in the ETINDE block located in the same area, whose oil volume, confirmed by EUROIL and SNH, was 155 million barrels. This suggests that the SW part of the RDR is still full of good reservoirs, notably R4, and therefore these results should in no way be ignored but rather taken into account for a much more in-depth study in order to judge the possibility that this reservoir is economically profitable.

Indeed, the volume of oil in place (STOIIP) in the R4 reservoir of the P2 segment is estimated at 28,000,000 sm3 or 176100628.3081 barrels of crude oil for a net height of 13.77 m TVD while the gas is estimated at 24 million m3 for a net height of 12.43 TVD, which at first glance seems enormous for a basin that produces in its entirety (55 fields) about 21.4 million barrels of oil per year (SNH, 2011). However, it is necessary to note that only 30 to 40% of these reserves are recoverable and in comparison with the recent discoveries made by the company ADDAX

PETROLEUM on the block IROKO fields in the offshore neighboring our study area, these reserves estimated at 20 million barrels of oil and 200 billion cubic feet of gas not including the additional reserves highlighted during the assessment phase with a net oil height of 38.6 m TVD and 65.1 m net height of gas We can note that this considerable difference in height and results obtained does not devalue the calculated hydrocarbon reserves of R4 because the latter presents a very large lateral extension over the entire study area, although limited in certain places by faults, and thus highlights the possibility that other reservoirs belonging to the deltaic alternations with greater potential may be the subject of further, much more in-depth studies in the future.

GENERAL CONCLUSION

The structuring of the Rio Del Rey Basin linked to the argillocinesis and the gravity tectonics work to reinforce at any time the quantity and types of traps today met in the different oil fields of the basin. New traps were created over the years and new migrations of hydrocarbons took place from the source rock (Akata) to the reservoirs (Agbada) or from reservoir to reservoir. The south-western part of the Rio Del Rey does not escape from it!

Faced with the considerable decline in hydrocarbon production of 13.08% compared to 2008. The study of this part of the basin was therefore to reassess its oil potential while showing that there are other fields outside those currently exploited and which tend naturally to depletion. We used seismic stratigraphy, qualitative and sequence analysis applied to logs as well as 3D geological modeling to calculate the volume of hydrocarbons contained in a defined reservoir. Several elements of answer could be brought within the framework of this work and constitute already a probable beginning for further studies more refined. We therefore distinguish several main objectives highlighting the potential of the block:

> traps: faulted anticlines, traps against faults

> place of accumulation: the western flank of the clay dome

> extension of the R4 reservoir: about 10 km from well P1 to well P2

> R4 reserve estimate: STOIIP = 176100628.3081 barrels of crude oil

We have shown that the R4 reservoir, taking into account the uncertainties, could be a deposit if it turns out that the volume of hydrocarbons determined is close to reality. This reservoir, like so many others in the SW block of the Rio Del Rey, must undergo further in-depth studies in terms of profitability calculations before considering the possibility of installing development wells.

Thus, the work carried out has covered the commitments in terms of evaluation of the potential of the SW part of the Rio Del Rey from two oil wells. However, the major problems encountered in the realization of this work are related to the amount of 3D seismic data to interpret. On the other hand, this internship carried out for a duration of 3 months was particularly beneficial to us because it allowed us to: discover 3D seismic, its functioning and its applications; have basic knowledge on PETREL software; understand the architecture of the SW block of the Rio Del Rey by identifying, pointing and mapping the clay dome zone; better understand the notion of 3D reservoir modeling; apply the knowledge of sequence stratigraphy to real well data; make a big step in the professional field by manipulating field data.

Finally, we can point out as a possible prospect, the drilling of new exploration wells at a depth of more than 1500 m, notably in the NGUTI turbidites, as the major part of the productive fields of the RDR is located in the clay-sand alternations which are situated between 1000 m and 1500 m depth.

BIBLIOGRAPHY

Benkhelil, J., et al (2002), Lithostratigraphic, geophysical and morpho-tectonic studies of the South Cameroon shelf, Marine and Petroleum Geology, 19, 499-517.

Blin, B. (1997). Kinematic, tectonic, sedimentary and petroleum synthesis of the Rio del Rey basin (Cameroon).

Bourquin, S., (1991). Facies-sequence analysis by logs of the Triassic of the central-western Paris Basin: Contributions to the reconstruction of the depositional environment. Thèse Doctorat université de Nancy I. 231 P.

Dluz, A., E. Mba and P. Montagnier, 1996. Geological and petroleum synthesis of the concessions

Ekundu and West Boa-bakassi. Elf Serepca, 341- DE N°6/105, ALD/EM/PM/rcn.

DR Mvondo Owono, F., 2010. Cenozoic surrection of West African passive margins based on two examples: the South Namibian Plateau and the Northern Cameroonian margin. Thesis of the University of Rennes1, France, 324.

Mascle, J. (1977). The Gulf of Guinea (South Atlantic): an example of Atlantic margin evolution in shear. Mem. Soc. Geol. Fr, N.S., n° 128, 104 p.

Reyre, D. (1984). Geological synthesis of Cameroon.

Mitchum R. M., and Vail P. R., 1977. Seismic stratigraphy and global changes of sea level, Part 7:

Seismic Stratigraphy Interpretation Procedure. Seismic Stratigraphy - Application to hydrocarbon exploration, Payton C. E. edition, *AAPG Brief 26*, 135-144.

Mitchum R. M., Vail P. R. and Sangree J. B., 1977. Seismic stratigraphy and global changes of sea level, Part 6: Stratigraphic Interpretation of Seismic Reflection Patterns in Depositional Sequences. Seismic Stratigraphy - Application to hydrocarbon exploration, Payton C. E. edition, AAPG Memoir 26, 117134.

Mitchum R. M., Vail P. R. and Thomson III S., 1977 b - Seismic stratigraphy and global changes of sea level, Part 2: The Depositional Sequence as a basic Unit for Stratigraphic Analysis. Seismic Stratigraphy - Application to hydrocarbon exploration, Payton C. E. edition, *AAPG Memoir 26*, 53-62.

Oberto Serra, Memoir 7 Pau 1985. Volume 2: Interpretation of logging data, Elf Serepca, 1996. Cameroon: 1996 annual exploration report, Exploration Division.

Perrodon, 1983. Geological synthesis of Cameroon.

Schiefelbein, C. F., et al. (2000), Geochimal comparison of crude oil along the South Atlantic margins. In: M.R. Mello and B.J. Katz, (eds), petroleum systems of South Atlantic margins, A.A.P.G., 73, 15-26.

Segesman FF, 1980. Well logging method.

Subra, A., Brousset, J.B. (1988). Petroleum potential of the turbidite terms of the Rio Del Rey (Cameroon) ELF SEREPCA, 341-DE N° 8/113, AS/JPB/rem.

SNH, 2009. Review books on offer in the Rio Del Rey

Printed by Books on Demand GmbH, Norderstedt / Germany